THE GRAZING
REVOLUTION

The Grazing Revolution

Second edition ISBN: 9798218059941
First edition published by TED Conferences, LLC in
November 2013
First printing of the second edition, 2023

Keywords: climate change, TED Talk, desertification,
grasslands, livestock, herds, biodiversity, soil carbon,
carbon sequestration, land restoration, wildlife habitat,
ecosystem management

About the Book

This book is a companion piece to Allan Savory's TED Talk, delivered in February 2013 in Long Beach, California. It was re-published in 2023 by the Savory Institute with Allan Savory's updates to the last chapter, "Scaling Up."

Watch Allan Savory's talk on TED.com and YouTube.

Contents

Preface

Today, a perfect storm is bearing down on us. It owes its genesis to a trifecta of factors that is overwhelming humanity's options for a viable future: Our global population, having surpassed 7 billion, is rising exponentially; on every continent outside the polar regions, a process known as desertification is destroying the land supporting that population; and Earth's climate is changing considerably faster than the scientific community had anticipated.[1]

Our society believes that some form of technology will fix climate change, which we typically attribute to greenhouse gases emitted from burning fossil fuels. Certainly technology can provide benign sources of energy to replace fossil fuels. But it will take more than just clean energy to tackle climate change. Our agricultural practices — how we produce and manage the crops and livestock that sustain us — are, I believe, equally to blame. Specifically, these practices — having already destroyed many past civilizations — are largely responsible for the desertification underway around the planet. One statistic bears this out: Each year, the earth loses (to erosion) 75 billion tons of soil, mostly from agricultural land,[2] which amounts to 10 tons for every human alive today. More alarming is the advancing rate of soil loss, which leads directly to desertification. And according to mounting scientific evidence, it appears that desertification itself is a major contributor to — and not merely a symptom of — climate change.

To give you an idea of the scale at which desertification is occurring, look at this view of the world, courtesy of NASA:

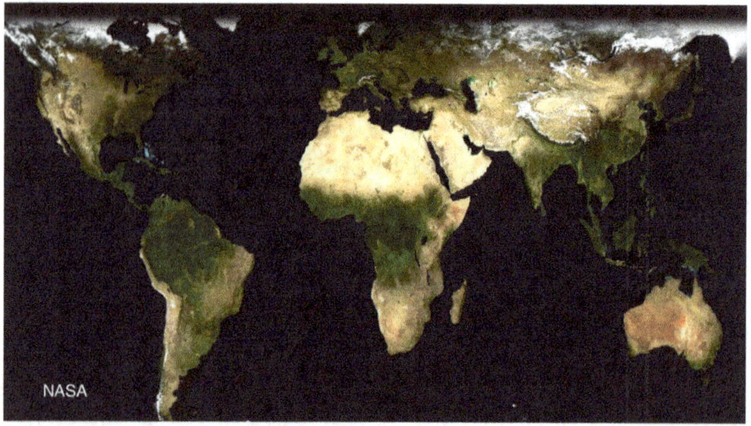

Generally, dark green land is not turning to desert, while the fringes of green outside the high latitudes and the brown are desertifying, even when rainfall is high. Keep in mind that the brown areas also encompass natural deserts, such as the Namib in Southern Africa and the Gobi of northern China, which get almost no rain at all, as well as parts of the Sahara.

Over millennia these deserts have been expanding into grasslands and savannas, resulting in failed civilizations. The deep black soils and abundant wildlife of Libya, described in the fifth century B.C. by the Greek historian Herodotus, are now part of the Sahara. The great civilizations of the Middle East — the Phoenicians, Persians, the 100 Dead Cities of Syria — all succumbed to desertification, as did the empires of northern China.[3]

Human-induced desertification happens when land becomes increasingly dry despite no change in rainfall. Rivers that once flowed year- round run only during periods of heavy precipitation and then quickly dry up. Silt and eroded soil fill dams until they're no longer capable of storing water. Aquifers are not replenished and water tables drop. Crop yields shrink. Grass and forage for animals, wild and domestic, becomes increasingly sparse. As a result, people who make their living from the land become impoverished. Unremitting poverty leads to social breakdown. Abuse of women and children rises, along with violence, as humans compete for scarce resources and access to land and water. The affected people forsake the land and swarm into cities or emigrate to other countries, creating a host of other social and economic problems.

The large area of brown on the NASA image — from North Africa, through the Middle East, across the former Soviet republics to China and into Pakistan and India — are suffering the symptoms of desertification. It is no surprise that these regions are the most volatile, violent, and unstable in the world. Desertification is equally severe over large swaths of the western United States, Mexico, Argentina, Chile, and Australia. It won't be long before nations are fighting wars over water — wars that will likely be greater and more violent than any fought over oil.

This vast environmental change began tens of thousands of years ago, when humans became predators. Research shows that humans had once been more akin to omnivorous scavengers, on equal footing with most other species. But when hominids mastered fire, evolved language and organizing skills, and developed weapons such as the spear, they became formidable predators. This was particularly the case in the grasslands where their prey ran in herds — grasslands whose deep, rich, water- and carbon-holding soils had developed over millions of years thanks to a balance of grazing animals, the predators that fed on them, and infrequent lightning-sparked fires (commonly associated with rain).

Predators hunting in large packs had to isolate a single animal in order to kill it, but humans could now wipe out entire herds of animals. They drove the herds over cliffs and into bogs, or used fire to corral and then slaughter them en masse. With this newfound advantage, humans began changing the environment around them drastically. Too few animals remained to graze on — and thereby spur the replenishment of — grasses and other plants, resulting in dead vegetation instead.

As the work of Tim Flannery has brilliantly explored in Australia, early humans across the globe used fire to eliminate this rank vegetation and bring on a flush of new grass to attract their preferred prey. They also used this method to provide green feed for livestock, which humans domesticated around 12,000 years ago. All of my research supports Flannery's assertion that such burning can expose soil in a way that makes it easily carried away by rain and wind. Without grass and soil-covering plant litter to retain it, residual moisture quickly evaporates. The world had never known fire in such frequency and expanse. And in the millennia that followed, vast human-made deserts formed and spread.

Early humans blamed desertification on livestock. They believed that livestock numbers were too high and this resulted in their overgrazing grasslands, leading to desertification. This conviction so permeated society that it assumed scientific validity and remained the explanation for desertification for more than two centuries. Nobody questioned this explanation — until recently. It's now time to examine the antiquated roots of these beliefs to see where the truth lies.

Chapter 1

A Radical Idea

The roots of my interest in land preservation began in early childhood. I was born in 1935, in Bulawayo, Zimbabwe (what was then Southern Rhodesia), and grew up with a civil engineer father whose dream was to construct the world's first mega-dam (which he ultimately had a hand in). When building roads, he went to extreme lengths to ensure they were scenic. Rather than felling an old tree or blasting through a large boulder, he'd route around them. I spent weekends as his schoolboy assistant helping survey dam sites in environmentally sensitive areas, something he would not entrust to his staff. We wanted to ensure these sites remained beautiful. And we spent long days in Zimbabwe's Wankie Game Reserve, now a famous national park, where across vast tracts of arid land my father built dams to provide water for wildlife. My father made it impossible for me not to love the bush — and by the time I left high school, I could not imagine spending my life anywhere else.

In 1955, after finishing my studies in ecology at the University of Natal in South Africa, I started working in earnest to save the desertifying land in my homeland (which had become part of the Central African Federation of Northern and Southern Rhodesia and Nyasaland). The task before me seemed simple. But during my first year in the Northern Rhodesian Game Department of the British Colonial Service, I began to realize that much of what I had been taught about ecology bore little relationship to what I was experiencing and observing in the field. Almost everywhere I looked, even in the wildest areas, where rainfall was plentiful and livestock absent, the land was deteriorating, or desertifying, as we would call it today.

I soon realized that desertification was not only more widespread than many people recognized, but it was closely linked to our agricultural practices, particularly those in the world's grasslands —broadly defined, grasslands include all environments in which grasses play a critical role in stabilizing soil, from dry deciduous forests to savanna-woodlands to open grasslands to arid and semiarid rangelands. Those practices were chiefly associated with livestock production, as only about 30 percent of the world's land is considered arable and suited to crop production.

Add to that the periodic burning of millions of hectares of tropical forest to create land for grazing, and you get devastating results:

- The soil carried down rivers and deposited over the shallow ocean shelves surrounding continents can damage rich zones of biological diversity — some of the most productive in the seas;
- The loss of our ability to store water in soils, which constitute the greatest available reservoir of nonfrozen fresh water (larger than all dams, rivers, and lakes combined);
- The loss of the soil's ability to store carbon, without which soil life is severely diminished.

When you combine these consequences with the appalling water and air pollution from industrial models of crop production and the unsustainable techniques used for factory farming pigs, poultry, and cattle, it becomes apparent that modern agriculture is a major contributor to desertification and climate change. By degrading soil, we diminish its tremendous ability to capture and contain carbon. This leads to the release of massive amounts of carbon annually from our earth's soils — a major and under-appreciated contributor to climate change. If we do not address the agricultural problem realistically and rapidly, climate change could continue long after we replace fossil fuels with cleaner energy sources. Investing in an agriculture that produces more food than eroding soil is far more critical to our survival than financing bank bailouts or the development of ever-more-lethal weapons.

There are people pushing for serious investment to explore the use of technology to geoengineer the Earth's climate, given the urgency of the task. If there is one lesson we should have learned, it is that technological "solutions" involving nature's complexity inevitably lead to unintended consequences, most of them adverse. Many examples come from agriculture: Chemical fertilizers to increase production have killed off microorganisms in soil, decreased soil fertility and water-holding capacity, and increased flooding; pesticides used to treat internal parasites in livestock have led to the destruction of dung beetles, which are vital to soil rejuvenation. To stake the future of humanity on using technology — in any form — to manage Earth's climate is as risky as playing Russian   roulette with cartridges in all six of the revolver's chambers.

Another idea often promoted to address climate change and desertification is the planting of trees. As the thinking goes, trees will both sequester, or store, carbon and reduce soil erosion. But in many parts of the world, trees won't solve either problem. Here's why:

- Desertification cannot be reversed by spending vast amounts of money and labor on the planting of trees and shrubs over billions of hectares of land where rainfall is only sufficient to sustain full soil cover under grasses. Additionally, in many cases, vast sums are spent on fossil-fuel-powered earth-moving techniques to harvest the water needed to keep the trees alive. It is not a commercially viable or scalable solution.

- As for the perception that trees can stabilize soils, create barriers for dust storms, and regreen desertified areas, planting trees where they do not naturally grow cannot address the reason why the land was desertifying; this approach treats the symptoms, not the cause, of desertification. Desertification will continue and many of these trees will die off.

Having spent five decades observing the problem of desertification — with countless hours around campfires discussing the issue with fellow scientists — I now stake my life on just one solution: Use livestock to mimic the formerly vast herds of wild grazing animals. My years of working in the field have made clear that large herbivores, properly managed, can reverse desertification on the scale required. These herbivores can be wild or domestic, but for practical reasons the job will fall mainly to cattle, sheep, goats, horses, and camels. They are the biological decaying mechanism that completes the cycle of birth, growth, death, and decay, thus maintaining the health and integrity of grasslands.

I realize this is a radical idea. For centuries we have believed that animals — particularly livestock — *cause* desertification. But my research has shown that livestock are the *only* tool now available to reverse desertification and return carbon and water to grassland soils. In the process, we will feed more people and enhance societal well-being.

This idea is counterintuitive to everything ecologists have been taught. So in this book, as in the TED Talk it is based on, I want to share my journey of re-education and discovery in the hope that it proves enlightening. Science supports this journey, which has resulted in a simple method any pastoralist can use to move massive amounts of carbon and water from the atmosphere back to the soil and begin reversing thousands of years of human-caused desertification. It is a journey that, above all, ends with great hope for the future.

Chapter 2

Early Mistakes

For centuries, nobody questioned that desertification was caused by too many animals grazing the land until nothing but bare soil remained. I, too, never questioned the dogma while studying ecology at university. As a result, I made some grievous errors on my journey to find the true cause of desertification.

The idea that desertification is caused by too many animals overgrazing plants is frankly as wrong as our one-time belief that the world is flat. But how I came to this conclusion is anything but straightforward. Rather, it was the result of an enormously varied set of circumstances, starting with my days in Northern Rhodesia's game department (there were no national parks at the time) tracking down man-eating lions, problem elephants and poachers, and later, in the Southern Rhodesian army, during a long and bitter civil war, tracking armed guerillas trying to leave no hint of their passage. Over the course of 20 years, I spent thousands of hours tracking animals and people over every type of landscape imaginable. Night after long night, lying in the bush alone, I reflected on my day's observations, trying to interpret what I was seeing. Why was the tracking easy yesterday over land with few livestock, but harder the day before over land where wildlife was plentiful but livestock were not? And why today was it easier tracking through tribal lands with abundant but scattered livestock? What was causing the differences when the soils and climate were the same?

I was brought up "knowing" that the terrible desertification in Africa was due to too many cattle, sheep, goats, and donkeys grazing the land. I saw the destruction first hand when surveying dam sites with my father in rural communities all over Southern Rhodesia's driest regions. My university training further entrenched my beliefs because I was taught that to keep grasslands from deteriorating, it was essential to burn them so they would replenish. But then, in 1956, as a young ecologist in the British Colonial Service in what is today Zambia, I had an experience that changed my entire way of thinking.

After graduating from university at age 20, I got a job in the Colonial Service's Northern Rhodesian (now Zambian) Game Department. I was tasked with burning vast areas of game reserves to ensure a fresh green flush of grass for the many game animals. But I soon had doubts about the wisdom of this practice. The areas we were burning were eroding terribly. Whenever it rained, I took walks in the rain to see what was happening to the burned land. I found that when rain hit soil surfaces made bare by fire, the water ran off in torrents, carrying massive amounts of soil with it. It was obvious that our burning policy was triggering tremendous erosion before grass had any chance to regrow.

In 1957, I wrote my first paper questioning our use of fire to encourage new grass growth. But my peers rejected the paper; they continued to believe that burning was essential to keep grasslands healthy. Nonetheless, I had discovered the first piece of the puzzle: Burning grass was keeping mature plants alive but hurting the underlying soil.

Around the same time, I was involved in managing wildlife areas slated to become national parks. In an effort to protect the land and wildlife, my colleagues and I relocated the hunting peoples who had lived on these lands for centuries (unfortunately, people were moved from their ancestral homes many times in colonial days; the belief at the time was that such things were justified by the greater cause of the nation having a national park). Yet, despite our efforts to provide full protection for the wildlife by removing the hunting peoples (and there being no livestock in such wild areas), within a short time the land began to deteriorate alarmingly, with some species of plants and animals disappearing altogether. When, in 1959, I transferred to the game department in my own country, today's Zimbabwe, I witnessed a similar plight: Areas set aside for future national parks were in demise. It seemed we were doing everything right but the land was still dying.

In the mid-1960s, I studied a severely deteriorating area in Zimbabwe where all the cattle had died in a series of dry years. Some 50,000 head of impala, zebra, wildebeest, and other game met the same fate. With almost no animals left on the land, I was sure it would quickly recover. It didn't.

In 1963, I published a paper in which I concluded that the land had been so damaged it could never recover. This paper passed the review process because my peers held similar beliefs. But we were all wrong; I still naturally assumed the deterioration in Zimbabwe's national parks, where there was only wildlife and no livestock (due to the tsetse fly, whose bite caused a fatal livestock disease), was also due to too many animals. Logically, or so it seemed to me at the time, the only thing left to blame was an over-abundance of wildlife, and the main offender appeared to be elephants. So I did more research, this time "proving" that too many elephants, buffalo, and a few other species were preventing these lands from recovering.

The consequent decision that elephant numbers would have to be reduced to a level the land could support was emotional and political dynamite in Zimbabwe. The government formed a committee of experts to evaluate my research and recommendations.

In the end, it agreed with my findings, and over the following decades the government culled some 40,000 elephants. But the land didn't improve and the damage still continues unabated in these national parks today.

The tragedy of my misunderstanding of desertification — and the subsequent reduction of the elephant population — had only one good outcome: Once I realized my horrific error, I became obsessed with finding solutions, devoting the rest of my life to doing whatever it would take to reverse the damage.

Chapter 3

Understanding the Cause

The Formation of Man-Made Deserts

Before we go any further, it is crucial to fully understand how the process of desertification works. Natural deserts, such as the Gobi in China and the Namib in Namibia, probably never were grasslands — at least not in the last few million years. There have always been natural, or true, deserts associated with regions of insignificant or no rainfall. These are not the landscapes I am referring to when I talk of desertification — a process of becoming or turning into desert because human management practices have made available precipitation less effective.

"Effective" precipitation is that which soaks into soil and only leaves it in two ways: through plants that transpire the moisture back to the atmosphere, or by flowing through soil into rivers, wetlands, springs, and underground reservoirs.

"Noneffective" precipitation is that which runs off the soil surface or evaporates directly out of the soil into the atmosphere.

The effectiveness of precipitation is determined by how much soil between plants is exposed. When raindrops hit the ground directly without vegetation or dead plant material — known as litter — to dissipate their energy, they break the soil particle structure apart at the surface, releasing organic components that get washed away. (The energy in a raindrop is dependent on its size and velocity. Large drops carry great energy, whether they're emanating from clouds or dripping off tree limbs. The spattering pounds the surface, breaking up the crumb structure and creating a near-impenetrable seal on the surface of many soil types that inhibits water infiltration and air exchange.)

More than any other factor, *bare soil surfaces result in precipitation becoming less effective*, and thus lead to desertification.

First and foremost, if rainfall is to be fully effective, it must penetrate the soil surface. How much water penetrates the surface is dependent on two things: the rate of application and the permeability of the top millimeter of soil. Two variables affect the rate of application: how hard it rains (precipitation intensity) and how quickly that rain flows across the soil (run-off rate). We can't do anything about precipitation intensity, but we can do a lot about the runoff rate. The faster water flows across the soil surface, the less time it has to soak in. However, if plant litter and closely spaced vegetation covers that surface, runoff is slowed and more water soaks into the soil, even during heavy downpours of rain.

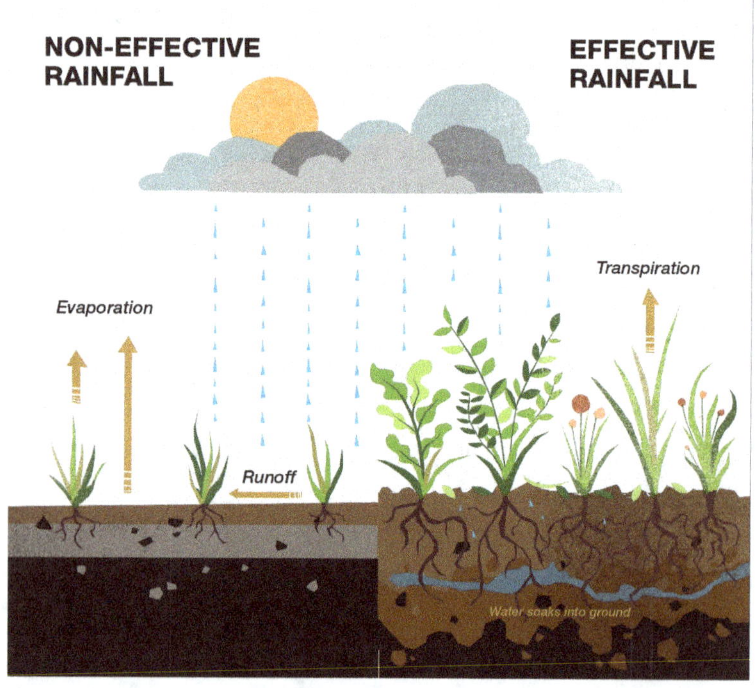

On the left, most rain flows over or evaporates out of the bare soil. On the right, most rain soaks into the soil and only leaves through plants or through soil to rivers and underground reservoirs.

During my walks in the rain in the early 1950s, I noticed that standing grass plants held banks of litter in place, forming millions of micro-dams. When I examined these micro-dams days after a big rain, I noticed they had briefly trapped water at the surface, which caused more rain to soak in above each dam. Now multiply this mechanism over millions of hectares.

Therefore, to ensure maximum precipitation effectiveness, it is essential that closely spaced plants and/ or litter cover the soil year-round. Doing so reduces both the frequency and severity of man-made floods and droughts. But more important, it maximizes the amount of water the soil can absorb during each rain event. And that is what I mean by precipitation "effectiveness," a factor usually far more important than the volume of rainfall.

Water loss through excessive runoff is common, and readily seen in many types of terrain and landscapes. As a result, a variety of techniques, some dating back to the Nabatean civilization of 200 B.C., have been developed to either reduce runoff or exploit it — for example, water-harvesting swales, keylining, and contour ridging. But in terms of water loss, runoff pales when compared with evaporation. During evaporation, water is literally sucked from exposed surface soil. Researchers in Namibia estimate that up to 83 percent of water that initially enters the soil during a rainstorm will evaporate almost immediately from the rangelands there.[4] Thirty years ago, in Pakistan, the scientist in charge of a forest research station showed me a project he was doing that involved introducing grasses from around the world in the arid Baluchistan province. He told me that after a good rain, the soil was wet to a depth of three feet. But within a day or two, the moisture disappeared. The introduced grasses died as well, which he could not understand.

What he didn't grasp was that visible grass wasn't enough. There still was too much soil exposure between plants. Grasses alone cannot prevent desertification. Litter — the trampled-down dead leaves and stems that make precipitation effective — is essential. Moreover, although it's not understood (or at least not reflected in any scientific papers I have read), soil's water-holding capacity is cumulative. Effective rainfall, which by definition remains in the soil, accumulates each year, with ever increasing water flowing through the soil to rivers, wetlands, and aquifers. Excessively exposed soil becomes more drought-prone year after year; land with covered soil becomes increasingly drought-resilient.

In 1960, while I was still a research ecologist in the game department, the governments in Zimbabwe (then Rhodesia), what would become Botswana, and South Africa were assisting drought-stricken farmers in the headwaters of the Limpopo River system. Meanwhile, the International Red Cross was collecting money for flood victims downriver, in Mozambique. The official line at the time was that the drought in the upper catchment was due to climate change — or as officials put it: "We just don't get the rain we used to get." I found myself the odd man out by noting that during the worst drought we'd had in the upper Limpopo area, we received the sixth-highest rainfall ever recorded. Land conditions upstream did nothing to mitigate runoff. Instead, the water surged downstream. So, in the same river system we had drought in the upper catchment and flooding at the lower end. *That is desertification in full swing.*

Flooding isn't always linked to desertification. Atmospheric humidity protects about a third of the world's land surface from desertification. These perennially humid environments appear dark green (or pale green in high-latitude regions) on the NASA satellite-view shown in the Preface. These regions do not desertify, no matter how badly the land is managed, because humidity fosters rampant plant growth. Vegetation covers bare soil quickly in these environments. That doesn't mean mismanagement won't damage these lands at all. When decreased biological diversity impacts plant vigor and soil health — reducing its capacity to absorb water — flooding and drought soon follow.

Most of the world's land surface, however, experiences erratic and seasonal atmospheric humidity. In these zones, even when rainfall is high, bare soil does not regreen rapidly — especially, I argue, when animals aren't there to break up sealed or crusted surfaces with their hooves, and to lay down litter. It is this erratic distribution of humidity — not the amount of precipitation — that is more critical to desertification.

The photos below, taken in Yemen's Tehama Desert during a storm and the next day, clearly show the loss of rainfall that can occur when the majority of the soil is exposed.

The Tehama Desert during a storm in which 1 inch of rain fell.

The same site in the Tehama Desert on the next day.

As you can see, some rainfall is running off immediately, creating small rivulets, and some is soaking into the soil. However, on the following day, this land was bone dry, with little sign that rain had fallen.

Approximately 27,000 gallons of water that fell on every acre (or 254,000 liters on every hectare) simply disappeared in less than 24 hours. I've observed that the Tehama Desert, at least in certain areas, is not a desert at all. Rather, it's a savanna turned to desert by human mismanagement. There are similar transformations underway across large swaths of the United States, Africa, China, Australia, and elsewhere.

Our common (and I'd say incorrect) belief, confirmed in much of the scientific literature, is that desertification occurs only in arid and semiarid regions. In fact, it also occurs in high-rainfall regions, where annual precipitation tops 30 inches (762 mm). Generally, even in these areas, there is a large percentage of bare ground between plants if the humidity is seasonal and interspersed with dry months. Unfortunately, people tend to look across grasslands and miss the gaps of barren soil. A grassland with 90 percent bare soil between plants appears as a sea of grass when viewing it from a distance, and especially from inside a vehicle moving at 60 mph. Unless you stop and peer down *into* grasslands, you generally cannot see the extent of the bare ground.

Desertification in grasslands also goes unrecognized if the vegetation appears "healthy," defined as having plenty of grasses of the "right" species — that is, noninvasive and deemed useful for one reason or another. It's common to find grasslands judged in good to excellent condition when in fact between 50 and 90 percent of the soil between plants is bare and eroding visibly.

This grassland, which gets 35 inches of annual rainfall, appears healthy when viewed horizontally.

Examining the surface more closely, the grassland reveals considerable bare ground between plants.

Zimbabwe was once one of the most conservation-conscious countries in the world. A rancher who won the much-coveted Natural Resources Board trophy for the best managed land asked me in the late 1960s to evaluate his property to confirm that it was indeed as good as the judges claimed it was. The ranch manager took me to the best land on the property and left me to take soil samples and make notes. A wonderful sea of waving grass of the "right" species surrounded me. But there was trouble below. More than 95 percent of the soil between the grasses was bare and eroding, and many plant species were disappearing, found only in remnant patches.

During the late 1960s, in a Namibian courtroom, I represented a rancher whose land was being expropriated by the government at what he believed was an unfair price. To assess the purchase, the government had compared his ranch with another one nearby that had been sold recently, and the rancher's property had fared poorly in the matchup. A well-known range scientist was representing the government. He described the higher-valued ranch as a "perfection of management," testifying to there being no soil erosion of any sort. Court was adjourned to give me time to examine the higher-valued ranch. I spent a day gathering data in what appeared to be a beautiful sea of grass. But I testified that more than 90 percent of the land was bare and eroding *between the plants*. I included pictures of grass roots protruding more than half an inch (1.25 cm) above ground — clear proof of soil loss, amounting to about 600 tons per hectare (242 tons per acre) over the time those grass roots had been exposed. My evidence carried the day and my rancher's property was revalued. The decision didn't make me any friends within the government. But it revealed to me how little attention was being paid to the soil surface, especially between the plants. One might see a sea of waving grass, but to determine its health, you have to get on the ground, part the blades, and examine how much soil is covered with litter compared with what is bare.

By now I had learned that you can't judge the ecological health of a grassland — or assess the severity of desertification — by looking across it. What I still hadn't figured out was how to prevent or reverse desertification.

Chapter 4

The Discovery

Unraveling the centuries-old mystery of what could be causing desertification was initially a case of groping in the dark and making many mistakes. In hindsight, it's easy to explain desertification. My work shows that three management practices lead to desertification. In order of importance, they are:

- Overresting soils and plants
- Overgrazing
- Burning

Overresting Soils and Plants

It took years for me to understand the primary causes of desertification. But of them, the hardest to identify, and the most critical, was overresting. Over-rest occurs when there is inadequate or infrequent grazing and trampling by animals, a process that keeps grasslands and their soils in a highly productive, resilient state. In 1979, when I first visited the United States, I observed some national parks desertifying as badly as any place in Africa. The damaging effects of overrest were obvious in America, and not something I could have discovered in Africa, where size of animal populations (wild and domestic) obscures the problem.

Chaco Culture National Historical Park in New Mexico was once home to thousands of native pueblo people.

After observing overresting damage in Chaco Culture National Historical Park and other national parks in New Mexico, Arizona, and Utah, I managed to locate research plots scattered around these and other states that had been designed to demonstrate how quickly the land improves when livestock are removed. In 1981, the U.S. Council on Environmental Quality produced a report, *Desertification of the United States*, that explained the thinking behind these plots: "'Improvident pasturage' or 'overgrazing,' as it has come to be known, has been the most potent desertification force, in terms of total acreage affected, within the United States."[5]

In the 1930s, the federal government, attempting to reverse the region's desertification, killed more than 250,000 sheep and goats belonging to the Navajo

Nation, according to the Southwest Indian Relief Council. Sure enough, with the sheep gone from the research plots, the overgrazing stopped and the plants did grow profusely over the first few years. But as years passed, the desertification of the Navajo lands appeared to have *increased.* Fifty years later, I saw the research plots turning to desert, despite having no livestock grazing within them for half a century.

These two photos, published in a paper on climate change, show the Jornada Research Station in New Mexico in 1961 (top) and 2002 (bottom), illustrating the dramatic change I had found everywhere animals were removed entirely.
Images: Courtesy of Jornada Experimental Range

Overresting led not only to the early death of grasses that had evolved with grazing animals but also to inadequate disturbance of the soil surface by animal hooves, which break up the soil so that new plants can grow and which trample down dead leaves and stems to provide the litter that makes precipitation effective. The focus on overgrazing obscured the problem of overresting.

Making matters worse was a sincere but misguided concern by many people to preserve the hard crusts that develop when nature covers the bare soil with algae and other simple life. They believed that these crusts held the soil in place, minimizing erosion and allowing time for grasses and other plants to establish.

Chaco Culture National Historical Park sign.

In Chaco Canyon and in wilderness areas in Utah, there are signs warning the public to not step off paths because they will do damage to the algal crusting. And yet, Chaco Canyon once supported an irrigation-based civilization, which would not have arisen if the rainfall had not been effective. It was not a desert. This suggests such regions had times of *less-effective*, rather than *less*, rainfall. Once more people settled on the land, wildlife herds kept their distance (there were no livestock) and the land overrested. The algal crusts that authorities — and my critics — want to protect have done nothing to prevent desertification. They are a symptom of desertification in seasonally humid environments.

The Roles of Overgrazing and Animal Behavior

Most of the world's seasonally humid environments have livestock on them but the land is still turning to desert. Could the land be overresting? The answer is linked to the role of overgrazing and animal behavior in these environments.

A fundamental inaccuracy common to discussions on overgrazing is to speak in terms of overgrazing the land. But animals can only graze or overgraze *plants*. When we refer to overgrazed land, we discount the fact that plants can be overgrazed, while the land, or soil, is overresting. This understanding constituted a major eureka moment for me in 1980, some time after I'd found an ally in André Voisin.

Voisin was a French scientist studying plant physiology in European pastures in humid environments.[6] He had discovered that overgrazing of plants bore no relationship to the number of animals on the land. Overgrazing happened when plants were exposed to a grazing animal for too many days, or re-exposed to a grazing animal too soon after previously being grazed — regardless of the number of animals. Overgrazing was a function of recovery time for plants, not quantity of animals.

Slowly the murky picture became clearer. The healthiest land I had seen was always associated with the largest herds — thousands of buffalo, elephants, and other grazing animals — accompanied by large packs of lions, wild dogs, and hyenas that kept them concentrated and needing to move off ground fouled by their own dung and urine. That movement minimized the overgrazing of plants.

Overgrazed plants result in a loss of soil-covering litter. And widely spaced animals trample very little litter onto the ground, whereas bunched animals trample more plant material to cover soil. Animals bunched and excited — by the presence of predators, for example — not only trample more plant litter to the ground but also chip the soil surface. The broken surface crust enables the soil to breathe, making it more permeable to water. This, in turn, improves conditions for the germination and establishment of new vegetation, leading to tighter plant spacing, which enables more litter to withstand the forces of wind and water and stay in place. It's an ecological feedback loop that grazing animals and their pack-hunting predators influence more than rainfall.

From my observations and Voisin's research, I concluded that with either wild animals or livestock, plants would be overgrazed if animals weren't moving adequately — either because they were no longer subject to predation or, in the case of livestock, because their movement was constrained by herders or fencing. The principle, it seemed, was that continually bunched animals led to greater soil cover between plants, closer plant spacing to hold litter in place, and more effective precipitation. And from what I knew about plant physiology, this would lead to more extensive root systems reaching to greater depths. Soil scientists, especially Sir Edward John Russell, who wrote *Soil Conditions and Plant Growth*,[7] established that the more extensive the grass roots, the greater the soil life, which, in the form of trillions of microorganisms, provides the nutrients plants need for growth. When grasses are overgrazed, their roots wither and the soil life and organic matter degrades to the point that the soil's ability to store water and carbon is measurably diminished.

Partial Rest — Overrest With Animals Present

With these principles in place, I finally understood why so much of the soil was bare between the plants in most of the world's seasonally humid environments. Where large herds of *bunched* grazing animals were replaced by scattered animals (wild or domestic), the land was suffering from *partial rest* — too few animals present, providing minimal vegetation and soil disturbance. In the absence of pack-hunting predators, animals scattered while grazing and trampled few or no plants, leaving the soil bare between them. In other cases, the animals overgrazed plants by lingering too long in a place, or returning to it too soon. In short, there was an imbalance — too much grazing or trampling in some areas and not enough in others. As a result, there was less forage to cover soil and feed animals.

This perennial grass plant has been insufficiently grazed and is oxidizing.

Scientists like myself, who had insisted on reducing animal numbers, had merely aggravated the problem. Fewer animals led to more ungrazed plants, which then accumulated dead, oxidizing leaves and stems that blocked sunlight from reaching new leaf buds and emerging stems at the plant base, eventually killing the plants.

A property boundary near Chaco Canyon shows the parallel destruction wrought by overgrazing (left) and overresting (right).

The photo above illustrates my assertion that over-resting the land is the single most damaging manage-ment practice in seasonally humid environments. On the left side of the fence, the desertification is blamed on overstocking and overgrazing. On the right side, there have been no livestock for more than 50 years.

The land on both sides of the fence is desertifying. On the right, total rest (no livestock); on the left, a high level of partial rest (a few wandering bands of sheep). No grazing on the right; overgrazing on the left. Over time, land responds to all of the influences on it. But in this case, the dominant influence — over-rest (total and partial) — made the greatest impact. Rainfall here is noneffective, making nearly every year a "drought" year.

What I have just described is vital to addressing climate change and altering the future of humanity. Prominent institutions such as the United Nations' Food and Agriculture Organization (FAO) have recognized the important role grasslands can play in mitigating climate change through absorbing carbon dioxide from our atmosphere. However, my research takes this a step further. We at the Savory Institute have shown that proper grassland management is essential to successfully address climate change. It's an extreme conclusion that has received much criticism over the years. My opponents claim that the carbon-sequestering capacity of grassland soils is not significant enough to affect such a global shift. Along with scientists like Christine Jones of Australia, who launched the Australian Soil Carbon Accreditation Scheme in 2007, I argue that such conclusions were drawn from measurements gathered from desertifying grasslands, where carbon already had been lost through oxidizing plants and loss of organic matter. In reality, grasslands not only play an important role in ambient carbon cycling, they also store carbon from millennia past. It's why many of the world's primary grain-growing regions — with their deep carbon-rich soils — are former grasslands.

Consequently, policy makers — and those using data from these various research efforts to build climate models — have been misled. Fortunately, a few soil scientists, led by Jones, have spotlighted such blunders and are now studying regenerating grasslands to expand our understanding of their potential to store carbon and water.

Burning

To rejuvenate overrested grasslands, many land managers burn them, a widespread practice still supported by many scientists and major environmental organizations. However, burning breaks down vegetation through rapid oxidation, adding black carbon, or soot, and other pollutants to the atmosphere. It exposes soil, and if done repeatedly, leads to a shift in grass species to those that are more fibrous and fire-dependent, and less nutritious. Animals, on the other hand, break down vegetation through biological decay and give soil much-needed disturbance.

While many scientists vilify trampling as destructive to plant life, my experience in the field has amply shown just the opposite to be the case. Trampling breaks hard, crusted soil surfaces, allowing new plants to sprout, while also pushing detritus down to the surface, where soil organisms can more easily break it down. The litter keeps the soil covered, and the dung and urine provide a natural fertilizer. The process makes the soil increasingly capable of absorbing and holding water and carbon.

Once livestock are returned to the land in proper numbers, their movement then managed with planned grazing and trampling, burning is generally no longer needed to keep grasslands alive. Planned grazing and trampling ensures that livestock don't graze or trample too long in one place, or return to it too soon.

A Few Words About Fire

For millennia, humans burned grasslands to promote green growth and attract the animals they hunted. In recent times, pastoralists, ranchers, range scientists, environmentalists, and ecologists have adopted burning to prevent grasslands from shifting to nongrass plants, such as weeds, brush, and trees. This practice is referred to as "prescribed burning." Using fire as a management tool is commonly justified by arguing that fire is natural. But no fire lit by a human is natural.

Fire *does* remove the dead plant material that blocks sunlight from reaching growth points at the grass plant's base, allowing plants to flush green early in the next growing season. However, because fire also exposes soil, if used repeatedly, this method leads to a dominance of grass species that mainly establish on bare ground, and thus to a largely fire-dependent grass community. Over and over again, I have seen that, in the long term, fire cannot replace biological decay, keep soil covered, or cycle nutrients effectively. Fire exposes soil, ruthlessly.

It's my view that grassland burning is a primary reason that available rainfall is becoming less effective, leading to increasingly severe and frequent floods and droughts.

Additionally, grassland burning accelerates the release of carbon into the atmosphere. Proponents of burning claim there is no consequent increase in atmospheric carbon dioxide. They'll tell you that the subsequently regrowing grass quickly takes up all the carbon. But this is not entirely accurate. The smoke (and thus carbon) from grassland fires remains in the atmosphere for many months, as pilots like myself experience every year in Africa. And subsequent growth is cumulatively decreased over the years because the available rainfall, due to soil exposure and erosion, is ever less effective.

The Difference Livestock, Properly Managed, Can Make

Now let's consider what we can do with properly managed livestock to eliminate overrest, burning, and overgrazing.

The following photo of the Dimbangombe Ranch in Zimbabwe illustrates a fairly typical grassland in the tropics following four months of regular rainfall and high humidity.

The subsequent photo is of exactly the same site one month after the rains have ended. Notice that the grass stems and leaves are now drying out, having already moved their food reserves to their massive root systems underground. Food reserves are stored out of harm's way from grazing. The reserves will sustain the plants over the non-growing season and be drawn on later to support new growth in the following growing season.

Lush grasses rustle at Dimbangombe Ranch in Zimbabwe after four wet months.

One month after the rains, the same grassland is at a pivotal moment in its health. Only under the right conditions will it continue to thrive.

For this grassland and its soil to remain healthy, almost all the dying mass of stems and leaves has to be removed over the next eight months. This ensures sunlight reaches new leaf buds and emerging stems at the plant base. If the dead leaves and stems are not removed, sunlight cannot reach the ground-level growth buds, and the plants will fail to thrive.

Livestock can be managed to facilitate this cycle very effectively, as the three images below illustrate.

The Herds of the Past

A day after cattle grazed this grassland (top), a dying mass of stems and leaves, along with dung, remains on the ground as litter (middle). After the rains begin (bottom), green leaves emerge in the trampled area.

Early explorers recorded massive numbers of large grazing animals in America and Africa. But these herds were not found in tropical forests. Instead, they ranged in the temperate and tropical grasslands and savannas of seasonally humid environments. In tropical forests, where most herbivores are insects, there is no defined season when a vast amount of plant material dies off. Large herbivores exist, but they tend to travel solo or in small family groups. There are large predators, too, but they're few in number and do not run in packs. Instead, they stalk or ambush their limited quarry. In such environments, desertification is not occurring, and land responds favorably when rested, or left to nature.

This does not hold true in seasonally humid environments, where over millennia, large numbers of grazing animals once existed, including in the high latitudes. Because ecological principles do not change over millennia, we can now understand how the deep soils of the American prairie developed. It wasn't through burning by early Americans. The soils evolved over millions of years through complex relationships with seasonal precipitation, grazing animals, pack-hunting predators, and natural (lightning-strike) fires.

Although these photos show a present-day herd, they're useful for understanding how animals and plants coexist, whether those plants are stimulated by grazing, as they were historically, or over-grazed due to changed animal behavior. To thrive despite threat from pack-hunting predators, grazing animals evolved either to be small, stealthy, and fleet-footed, or to move in herds. Generally, the larger the herd, the safer the individual animal.

Wolves take down a lone bison.
Image: Courtesy of Jim Brandenburg

Wildebeest and zebra stick close together for safety.

There are many written records of the vast herds of bison, pronghorn, elk, and deer in the United States — herds taking days, not hours, to pass pioneer wagons. In Southern Africa, there also are accounts of vast game herds, such as this description by George Mossop, writing in the late 19th century:

The scene that met my eyes the next morning is beyond my power to describe. Game, game everywhere, as far as the eye could see — all on the move grazing. The game did not appear to be moving: the impression I received was that the earth was doing so, carrying the game with it — they were in such vast numbers, moving slowly and steadily, their heads all in one direction against the wind...Hundreds of thousands of blesbok, springbok, wildebeest, and many others were all around us.[8]

Chapter 5

Implementing the Discovery

Properly Managed Livestock

In the late 1950s, while I was an ecologist in the Northern Rhodesian Game Department, I struggled — with very limited staff and money — to save tens of thousands of a heavily poached antelope, called the black lechwe, in the vast Bangweulu swamps of Northern Rhodesia. As a possible solution, I came up with an idea I called "game ranching." The plan was to harvest the lechwe population in a controlled manner, thereby ensuring their survival in high numbers. We would harvest the lechwe just as we would a fish species in the ocean, culling the animals by sex to maintain a steady population. And to assist in the harvesting, we'd enlist former poachers, and we would sell the sustainably harvested meat at a fair price to their former customers.

I left the Northern Rhodesian Game Department in 1958 before I could execute my plan. But in the early 1960s, while working in the Southern Rhodesian Game Department, I was able to try it with two American Fulbright scholars, Archie Mossman and Ray Dasmann, who were pursuing the same idea, which Mossman had come up with much earlier in the U.S. Between us, we finally got the game ranching industry started in what became Zimbabwe. I was focused on desertification and believed that if we replaced livestock with managed wild populations of large herbivores, we could address the land degradation — the kind that resulted in increasingly severe droughts and floods despite no change in our rainfall — that was occurring throughout the country. Gradually, against great resistance from the game department (as well as from the veterinary department, which was aligned with the livestock producers who feared game meat competing with the beef they produced), we got the game ranching industry going. Later, it would spread to South Africa and Texas, where it became game farming with fenced-in animals much like livestock. In Rhodesia, rather than put up fences to contain the wildlife, at our instigation, ranchers began amalgamating ranches to allow more freedom of movement in order to utilize the game more effectively.

We were certain that this would lead to land recovery. *But the land continued to deteriorate*, just as it had in the national parks, and just as it's still doing on the ranches I have visited in Africa and Texas, where game farming is big business.

Our problem? The pack-hunting predators were not sufficient to bunch the wild herds and induce adequate movement to get enough hooves chipping and trampling the soil and grass; fences constrained the required movement as well. At the time, my antagonism toward livestock was well known in Zimbabwe because of the publicity I'd received promoting game ranching as a possible solution to land degradation. So I was surprised when, in the mid-1960s, an elderly ranching couple, Edward and Evelyn Rushmore, called on me for help. Their story was simple: They loved their ranch, had followed the advice of government advisors and the local research station, but their land was steadily getting worse. Would I help them? I agreed but on one condition: that they accept that I had no answers (yet). It would be the blind leading the blind.

Thanks to the Rushmores, I had a chance to experiment with livestock, to see if we could manage them in a way that didn't damage the land. I had spent several years reading all the wildlife and range management scientific literature I could lay my hands on, searching for clues that might help us manage in a way that didn't damage the land, but to no avail. Again, I turned to French pasture specialist André Voisin's work — as I was now dealing with livestock rather than wildlife alone — and it was there I found the vital clue about time and overgrazing, namely that plant recovery time was more important to healthy grassland than the amount of animals grazing on it. Soon after, I came across an article in a farmer's magazine by John Acocks, a botanist who was mapping the changes in vegetation occurring as desertification advanced across the Karoo region of South Africa. What caught my attention was Acocks' statement that South Africa was "overgrazed, but understocked" (stocking refers to the number of animals placed on the land) — a notion ridiculed by range scientists. I decided to visit Acocks and the ranchers he was helping, and I made an exciting discovery.

On one of the ranches, which belonged to a couple named Len and Denise Howell, I spotted a small patch of ground that appeared remarkably different from everything around it. It turned out that sheep had bunched on the spot for a short time during a terrible storm several months earlier. As a result, litter cover was heavier and the plants were dense and green. In a flash, I made a connection to what I had observed years before: healthier land that had been trampled by enormous buffalo herds kept bunched and moving by prides of lions. Soils, plants, and animals complemented each other and, in fact, needed each other to thrive. Immediately, I saw that livestock could have the same beneficial effect as wild intact game populations. (The Howells could not understand my excitement; they and Acocks were focused on preventing overgrazing of plants, while I was seeking ways of keeping soil covered.)

Livestock weren't the problem; it was how they were managed. If we could simulate the behavior of wild herds under threat of predation, we could use livestock as tools for land restoration.

A Holistic Grazing Planning Process

Beginning with the Rushmores (who were soon joined by a neighboring rancher, Colin Bickle), I tried various combinations of Acocks' grazing methods and Voisin's technique for eliminating overgrazing. But I fell on my face again and again. At first, livestock didn't fare well. They weren't breeding effectively and didn't gain enough weight. Wildlife also suffered. Their offspring had too little cover and feed at critical times. We had allowed livestock to graze too heavily. I also realized we had to plan for inevitable drier years, but I was not happy with the conventional solution to this problem, which was to reserve grazing areas. This gave the animals less grazing acreage, undermining nutrition. Plus, the forage in the reserved areas grew rank, and the soil and dead vegetation didn't get trampled, which decreased its productivity.

Remarkably, the ranchers stuck with me — even continued to pay me— knowing we had problems to solve. And I was desperate to solve these problems. But how? I knew centuries of pastoralists herding their animals had contributed to the formation of the world's great man-made deserts. Voisin had studied rotational grazing, in which livestock are moved through paddocks after relatively short periods of grazing to try to prevent overgrazing. Such practices had apparently first developed in Europe in the 18th century and had been exhaustively studied by Voisin, who pointed out that people practicing rotational grazing were routinely experiencing a loss of biodiversity having failed to understand that the recovery period for plants between the grazings was more important than the actual grazing periods. He had developed a simple planning process that ensured close attention to recovery times, calling it "rational" grazing.

Like Voisin, I knew we would have to develop something entirely new and innovative. Some form of planning process was needed for Africa's more complex — because of erratic humidity and many more species — grasslands. European pastures are in regions of almost perennial humidity. Dead grass material does not oxidize and cause the death of plants, and even very high levels of overgrazing do not lead to bare ground. Voisin's "rational grazing" method was not wrong, but it was not addressing the complexity I faced.

Somehow, we needed to develop a simple way of planning that dealt with the issues ranchers faced daily: varying and erratic seasons; challenges with making rainfall effective; livestock health and reproduction issues; wildlife that had its own needs; human families that had theirs; livestock that had to mimic wild herds under threat of predation; and so on.

Knowing that ecologists, wildlife managers, and range scientists had never attempted to plan holistically for anything even remotely as complex as this, I turned to other professions to see if anyone had.

I reviewed the planning procedures developed in various business schools but found them too theoretical, academic, or impractical. I finally found what I was looking for in Britain's Royal Military Academy Sandhurst. It had developed techniques for handling rapidly changing, complicated situations on the battlefield. Sandhurst's methods were practical, easily taught to untrained civilians in times of war, and effective when used by exhausted, often confused officers, who had imperfect knowledge of ever-changing situations. I latched onto Sandhurst's aide memoire (memory aid). It was a list of steps to remind soldiers how to devise the best possible plan in any circumstance. The trick was to focus on a single piece of a confusing situation, quickly consider all its details, and then focus on the next piece until the best possible solution at that time emerged.

Having found a planning process based on centuries of practical experience, there was one more hurdle to cross. In this scenario of battlefield planning, the armies of Europe had faced many variables but always over short periods. Ranchers had to deal with just as many variables but over much longer periods. And ranchers had to manage large areas of land that could change dramatically from season to season. Armies didn't have to handle animal behavior, or calculate feed volume over time, or any of the other factors that affected ranchers. So, I developed a 12-step aide-memoire for grazing planning that took all this into account, enabling any rancher to develop the best possible plan at a given moment. I also produced an accompanying grazing planning chart which recorded the results of each step and enabled the rancher to track every variable separately. The rancher then followed the plan, monitoring and adjusting for slight changes, such as a misjudged plant-growth rate (with the chart showing the effect of those changes on the important recovery period). The chart also eased replanning if something unexpected — say, a fire or an invasion of locusts — threw things off course. Thus was born what has become known as the holistic planned grazing process today.

By the late 1960s, I was working closely with scores of farmers and ranchers in five southern African nations (Rhodesia, Botswana, Namibia, Swaziland, and South Africa). For the most part, we were ignoring the prevailing "wisdom" — government and university officials telling us to decrease animal numbers and increase feed. George Rudland, the minister of agriculture in Rhodesia, was particularly recalcitrant. At one point, he challenged me to prove that ranchers could double their animal stocking rate using planned grazing and make more profit while not damaging the land. Rudland's senior advisors had told him this would most certainly lead to financial disaster and serious deterioration of the land. I accepted the challenge. The project was dubbed the Charter Grazing Trials.

We agreed that if I was proved wrong, I would close down my consulting business and stop "misleading the country" (as Rudland put it). But if planned grazing worked, Rudland agreed to instruct agricultural colleges and government advisory services to promote my technique. Charter Estates, a prominent ranch in Rhodesia, volunteered the land and the cattle. We selected paddock and water arrangements for two planned grazing areas and immediately doubled the animal numbers. These were compared with a control area following government and range scientist's advice with cattle at the "recommended" level to prevent overgrazing. As we proceeded, it became clear that the cattle in the government areas were gaining more individual weight than the cattle on the planned areas, which were present in larger numbers.

While this concerned the ranchers and range scientists who believed "cattle performance" — that is, individual cow weight — is the most important factor in achieving a profit, it did not concern me. The trial was not testing individual cow weight. It was testing my argument that running more animals would result both in greater profit overall and in improved conditions for the land (a key factor in long-term, sustained profit).

Although individual animal weight is important, it's not as important for profit as weight of meat produced per acre. Simply put, fewer slightly heavier animals do not result in as much meat produced as a higher number of slightly smaller animals. This is why so many ranchers who pride themselves on their cattle's conception rates and weight gains still go broke, unlike, say, corn farmers who focus on yield per acre and not the size of their corn cobs or the number of cobs per plant. Some agricultural economists argue that a higher gross margin per cow is more important to eventually covering overhead costs, and such considerations are definitely looked at when making management decisions holistically. But in the end, land is our most finite food-producing resource. And no matter what form of animal production provides the best gross margin, it still becomes more profitable to run more animals on the land. This is particularly true when running more animals also leads to sustained production because those animals are speeding regeneration of the soil.

And indeed, in these trials, the two planned grazing areas were more profitable than the government control area. The land in the planned areas had more leafy plants, more ground-covering litter, and in general, more and denser growth — all reflecting greater productivity. At the time, however, it was customary to evaluate the condition of the range instead by the number of species present (some species of grass being considered more desirable than others) and by basal cover — that is, the amount of the soil surface covered by plant bases, where the stems emerge, as opposed to overhanging leaves and stems higher up the plant. By these criteria, which were used by the government range scientist on the project, our plots and the government plots did not show much difference. However, in the final report, the government range scientist acknowledged the unmeasured difference favorable to the planned grazing areas, noting that the government controls were showing dead, oxidizing plant material that would require burning. Where the grazing was planned, he reported no buildup of dead material, and this "gave the sward a very healthy appearance."

Many critics have since continued to claim that these trials failed because the individual animals were heavier in the government plots — a false priority, already addressed, and actually, a problem we later solved as we gained experience with planned grazing. In particular, in the early days I was misled by research that showed that animals determined what plants to eat by selecting desirable species. Only after more research and increased observation in five countries did I come to understand that animals don't select species to eat. Rather, they select a balanced diet of protein, energy, and fiber or other nutrients from many species. This results in some species being favored at certain times of the year or during certain stages of the plants' growth. Once we understood this, we brought it into the grazing planning process to ensure that animals were never forced to graze heavily unless we deliberately wanted them to lose weight. This meant they were not limited in their ability to select a balanced diet, regardless of species, which in turn enabled them to gain weight.

Other critics pointed to the higher rainfall the area received in the years of the Charter Grazing Trials, ostensibly leading to better forage and therefore better results. However, the government plots benefitted from the same rainfall as the planned grazing areas. And in the years since, my best results with individual cattle performance always occurred in years of lower rainfall, something acknowledged by many ranchers.

The most dramatic case I recall was that of two ranching brothers I was helping in the south of Zimbabwe. Ben and Boet van Vuuren had four ranches close to a ranch of my own, and I had just started them off on holistically planned grazing when we were hit by a terribly dry year. It coincided with a plague of army worms (caterpillars that devour everything). Large numbers of livestock were simply dying of starvation, with nowhere to send them for grass, and ranches all around were frantically trying to unload their cattle at markets but failing. The van Vuuren brothers planned to scatter their cattle, but I asked them to let me try something new. I had the brothers amalgamate all cattle from the four ranches into one herd. Then, we created a grazing plan using the paddocks on all four ranches. The cattle were sometimes moved every day or even every half day, and they got desperately thin from just eating twigs (mainly from Mopane trees). But we did not lose a single animal. The ranchers who scattered their cattle lost thousands to starvation.

If the Charter Grazing Trials had truly failed, I would have been out of work (honoring my part of the agreement). On the contrary, the agricultural economist monitoring the results of the trial was so convinced by its outcome that he left his university position to join my consulting business, eventually forming his own company called Grazing for Profit (though in his consulting, he advocated lower stocking rates than the doubling I did). The manager of the ranch where the trials took place was equally wowed. He publicly admitted that he would be practicing planned grazing in the future, and he subsequently did over parts of the large ranch.

Holistic Planned Grazing in Brief

The holistic planned grazing process begins by clearly defining the context for the many objectives in management. In this case, it is a holistic context. This is opposed to the context we find for objectives in management, most commonly need, desire, profit, or addressing a problem. These contexts are too simplistic and don't take into account the social, environmental, and economic complexity involved.

A great many factors are considered and the details noted by those doing the planning. What land is available? (For instance, is it owned or leased, and what is owed on it?) Is extra land available in an emergency, and at what cost? What animals are there? What's the best way to divide the land into grazing areas, or paddocks? Should these areas be fenced or merely demarcated by landmarks known to herders? Where will the animals drink, and how can you ensure water supplies are adequate? What weather patterns should you be aware of? (Are there periods of heavy frost? Areas of deep snow? Prevailing winds? Fire threat?) Are there market factors that could influence the type of livestock run? (Year-round cows and calves, or seasonal steers on the gain, a mix of cattle and sheep, sheep and goats?) What specific wildlife needs and considerations have to be incorporated into the plan? (For instance, where do ground-nesting birds nest, what degree of cover is essential for chick survival, and over what months?) What other land uses — crop fields, forestry operations, hunting, ecotourism, game viewing — will affect where and when the livestock can be?

The first thing to consider is what season you are planning for: growing or nongrowing. You create one plan for the growing season, which aims to maximize the amount of forage produced while preventing over-grazing. You create another plan for the nongrowing season, where the aim is to ration out the available forage so it lasts until the next growing season for both livestock and wildlife, while ensuring enough litter remains to make the coming rains effective.

The next step is to set up the grazing planning chart. Across the top of the chart, note the months.

Down the left column, chart the grazing areas, or pad-docks, and rate them based on their quality and size. Livestock details, including quantity, are recorded on the bottom.

Next, use color coding to demarcate the factors that will influence your plan. Examples include when you expect to breed and wean; when and where areas will be snow covered or at risk of fire; when and where, say, antelope are dropping young and cattle will need to be kept away; when and where you will need herds to trample a bare ground area or a harvested crop field; locations of poisonous plants and times of danger (not all poisonous plants are always dangerous; also, the danger lessens when their leaves constitute a small percentage of total forage available to grazing animals); where there may be water restrictions; and when to expect absences of key people due to vacation, family events, or other commitments. The more informed and rich your plan, the more effective the management, and the better the overall outcome.

Your plan as it develops on the planning chart will eventually provide a clear picture of where livestock need to be and when. But before you can plan their moves, you have to determine how much recovery time plants will need between grazings. During the growing season, bunched perennial grasses generally require 30 to 90 days to recover from a grazing. In the nongrowing season, since plants will be dormant and unlikely to be overgrazed, it's more important to simply give animals the best available diet during this hard dry period. With the recovery period determined, you can then plan grazing periods for each paddock. Aim to keep them short enough to minimize overgrazing in the growing season; you want enough forage left behind to sustain animals during the nongrowing season.

Rancher Margaret "Mimi" Hillenbrand displays the holistic land plan and holistic grazing plan for her 28,000-acre ranch in Hermosa, S.D. She raises about 2,000 head of bison for grass-fed meat and the sale of genetically pure bison to other producers and conservation organizations.

Image: Courtesy of Kirk Gadzia.

Now you're ready to plan the actual moves of the livestock. All the critical events, such as calving times, are marked on the chart. Place the herd in the best paddock for these events to take place. Just as

complicated construction projects are planned back-
ward, from day of opening, you'll need to do the same
with some paddocks if they're already booked for critical events. Mark each animal move in pencil, because
juggling is often required. The aim, always, is to get
the animals to the right place, at the right time, for
the right reasons.

With the moves now planned, you should double-
check whether you will have adequate forage to feed
the animals, or whether you can run more animals
while continuing to meet the needs of other land
users, such as wildlife.

You're now ready to implement the plan. But mon-
itor it closely and modify as needed, because nothing
ever goes exactly to plan.

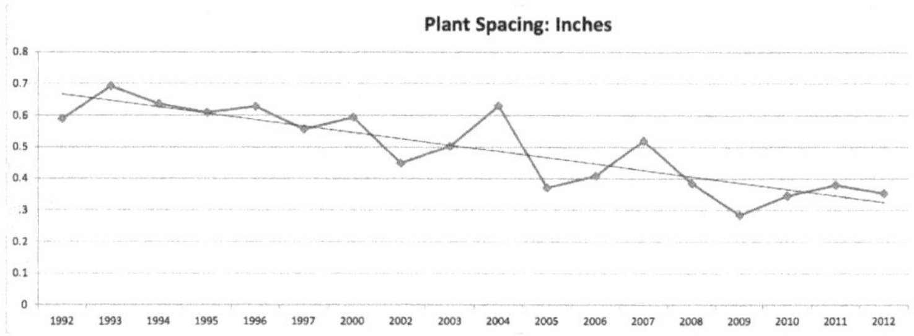

Plant coverage over the land has improved steadily at Mimi
Hillenbrand's ranch since she implemented Holistic Management in
1992. The straight black line represents the trend toward smaller
gaps between plants.

This brief outline of holistic planned grazing illustrates a process that has been under development for
nearly 50 years, with input from hundreds of people
— ranchers, farmers and the advisers assisting them,

pastoralists, and scientists hailing from nearly every continent and environment. Every problem or novel situation has helped us improve the process. Today, one aide memoire and one grazing chart can be used worldwide in commercial situations, with a slightly modified version for pastoral situations, where livestock are herded, literacy may be low, and planning is a community endeavor. This effort is now coordinated through the Savory Institute and the Savory Hubs it is establishing around the world (see the "Scaling up" chapter).

Are Livestock Appropriate for the Context?

With the holistic context clearly defined, we must decide whether or not running livestock is appropriate within that context in the first place. In some cases — for instance, if the decision is whether to clear tropical forest to run livestock — it's likely not appropriate. When evaluated in a holistic context, this might lead to short-term profit, but over the long term, it would diminish biodiversity, decrease the resilience of the environment, lead to social problems for displaced peoples, and eventually result in economic losses due to declining productivity. In this scenario, other options would need to be investigated that would regenerate the environment and the economy while keeping the management focus on what is truly valued now and long-term.

In other environments, running livestock may

initially appear to be inappropriate but later seem necessary. This looked like the case during a series of training workshops I led for the U.S. Department of Agriculture in the mid-1980s near the Sonoran Desert in Arizona. Participants were convinced that livestock trampling was contributing to the demise of the endangered desert tortoise. The solution, they insisted, was to remove the livestock and "leave it to nature" to restore tortoise populations. But rather than rely on such reports, I had participants go out in the field and make judgments for themselves. And we saw indications of something else entirely: The tortoises weren't dying out because of livestock trampling or cars running over them or even ravens eating them, as some people had also suggested. In fact, the land and plants were overrested. The too few livestock in the area simply could not have the same favorable impact on the land that former populations of pronghorns and wolves had once had. As a result, much of the soil was bare and there were many old, gray, and dying grass plants that had not been grazed or trampled in years. No new grass plants had been able to establish on the bare soil, and the dying plants were no longer producing seed. This is typical in such overrested sites, as we have observed on the long-rested research plots and severe desertification in some American national parks devoid of large animals.

The tortoises were dying out from what the great wildlife ecologist Aldo Leopold referred to as "welfare factors" — lack of adequate cover, feed, and moisture at critical stages in their lives. When the workshop group went out into the field, we saw this with our own eyes. Young tortoises, especially, had almost no cover to protect them from predators, and no small, green plant shoots for moisture or food — only inedible algal soil crust, scattered gray and oxidizing grass plants (themselves dying), and that was about it. Some participants reported tortoises feeding on fresh cattle dung for moisture and food.

The solution, we realized, was more livestock, not less — as long as they were managed with holistic planned grazing to mimic the very high numbers of pronghorn and other animals that once coexisted with the tortoises and plants (and the predators that kept the pronghorns bunched and moving). The holistic context encompasses the quality of life desired by the people involved and what they need to produce to attain it. It also encompasses a description of that land, not in its present state but as it will need to be far in the future to sustain their descendants and all other life forms.

If nothing but livestock is appropriate for this context, then we know there is no option but to run them or accept the tortoise dying out.

Chapter 6

Successes, Lessons, and Results

The ranchers I worked with initially in developing what has become Holistic Planned Grazing began to see tangible recovery of land, wildlife habitat, and profitability immediately. Forage production increased: Plants were no longer being overgrazed, more plants were establishing, and disappearing species were returning. Ranchers had the happy problem of needing to increase livestock numbers to prevent all the extra grass from overresting.

When word got out that ranchers were *increasing* livestock numbers, demand for my services spread like wildfire. The owners of an enormous — and seriously desertifying — 1.25 million-acre (506,000-hectare) ranch approached me for help. I wanted to test a new radial fencing layout, which incorporated a central watering point to bring costs down. Many condemned this layout, including Voisin. They feared it would lead to over-trampling at the center, where paddocks converged. I disagreed, positing that the animals would spend about the same amount of time on the land near the water as they did in areas far from the water, and, in any case, time on the land, not the number of animals on it, was the vital issue. The owners agreed to try it.

Established in the late 1960s on the Liebig Company's ranch in Zimbabwe, we called our test the Advanced Project. We chose our name carefully because we were increasing animal numbers to ridiculous levels. Had we called it a "trial" and failed, it would have done damage in the public eye.

Because I did not want the Liebig's owners to take any risk, we selected the worst land available: 4,000 acres (about 1,620 hectares) on which there was no grass at all, only scattered trees, shrubs, and weeds to sustain the animals. Sable and roan antelope, and other protected species, had once been abundant but had since died out due to habitat destruction. The ranch headquarters — originally sited in this drier southern country because it had been the most productive despite less rainfall (see below for an explanation) — had moved many miles north, to less degraded land. I was given full rein to do what I wanted. In addition to the new fencing layout, I intended to increase numbers rapidly on this severely desertified land to see if that would restore the land faster. I was intentionally trying to get the planning process to break down. If it didn't, then we would know that it was safe to continue to push up the animal numbers and try our planned grazing over the entire ranch and its 60,000 cattle. For most overcapitalized ranchers, being able to run more livestock without having to buy more land would turn losses to profits quickly. And by my reasoning, higher animal numbers mimicking nature would improve the land faster than lower numbers.

We proceeded, dividing the land into units, with radial fencing from a central point, where we provided water and a dip (to prevent tick-borne diseases). Despite the appalling state of the land and low, erratic rainfall that averaged 12 inches (about 300 mm) annually (but falling mostly in a two- to four-month period), we doubled the livestock numbers immediately. This proved too little and too easy, so by the end of the first year, we tripled the cattle numbers, far surpassing what anyone believed land in such bad condition could support.

We used the radial fencing to maximize the physical impact of the cattle, concentrating them as they moved through each paddock or grazing area. We didn't reseed the grass or provide the cattle with feed. We relied solely on the cattle to grow the grass by loosening the soil with their hooves, which let new plants emerge, then trampling down uneaten plant material to produce water-holding mulch. (See "The Recovery Process" sidebar below.) By the end of the first rains, grasses were returning.

Early in our project, a heavy rainstorm caused the cattle to concentrate in one area where they trampled nearly every plant. The staff panicked and urged me to let the cattle spread. But this wouldn't have been natural. Wild herds never scattered during storms, much in the way sheep and cattle don't in a snowstorm. So I suggested we leave them as they were to see what happened. Within a year, that very heavily trampled area was solid grassland.

Over the eight years that followed, the land continued developing into ever-denser grassland, regardless of rainfall. We compared the production of meat per acre on this land with the production on the surrounding 200,000 acres under conventional stocking rates and grazing practices, also owned by the Liebig Company. The result was astonishing: The land with greatly increased animal numbers under planned grazing produced five times more meat per acre than the surrounding 200,000 acres, which were continuing to desertify.

The Advanced Project infrastructure (fencing and water) cost a total of $1.80 per acre ($4.40 per hectare), a very small investment to increase production five fold while decreasing costs of production, which it did. Having gained confidence from this demonstration — and reassurance from once-doubters, including some of the ranch staff — we implemented planned grazing over the entire ranch of over a million acres.

The land on Liebig's ranch as it appeared at the start of the Advanced Project.

The land as it appeared three years later.

Next, we initiated a second Advanced Project (on a ranch belonging to another client of mine) 400 miles to the north of the Liebig ranch, in a high-rainfall area that averaged 40 inches (about 1,000 mm) annually. The rancher who owned it wanted to know whether we could achieve the same cattle and land performance by pushing animal numbers rapidly in this higher-rainfall, long-grass country. (In higher-rainfall areas, the forage is never as nutritious as in lower-rainfall areas, primarily because the greater amount of precipitation causes soil minerals to soak down or "leach" to lower depths, beyond the roots of the grass plants, which are less nutritious as a result.) And again we achieved similar results. I began working with ranchers in five countries in Southern Africa and eventually helped landowners on four continents practice holistic planned grazing.

The Liebig Advanced Project was conducted using methods so new and counterintuitive that they were widely condemned at the time by academic authorities throughout Southern Africa. My preferred yardstick was the health of the land and the feedback of the owners of the ranch. The principals at the Liebig Company were so impressed with the outcome that, as mentioned, they adopted planned grazing across the entire property of over a million acres, and with ever-increasing success. They also asked me to introduce the same planning to their ranches in Paraguay. Years later, critics in America continued to claim the project was a failure, and some still do to this day.

`The Recovery Process

If animals are present, they are eating something: weeds, twigs, fallen leaves. And they are probably wandering over a large area to find this food. Once you divide that larger land area into smaller pieces and combine the dispersed animals into a single herd that grazes each piece briefly, plants that were being grazed day after day are now given a chance to regrow. Similarly, if I were to shave my face daily, very little hair would accumulate between shavings. If I were to shave every two months, a lot more hair would grow between shavings. If animals graze the same plants every day, the plants don't have enough time to grow new leaf before being grazed again. But if those same animals are only allowed to graze an area for short periods interspersed with abundant time for plant recovery, any growing plants will produce a lot of leaf before they're grazed again.

This is all sorted out in the holistic decision-making and grazing planning. While animals graze in the planned area, which can be fenced or unfenced and herded within, the plants on the vast majority of the land begin to recover. Because the grazing period is short in each area grazed, new perennial grasses can establish without the threat of immediate overgrazing.

There's more to the story. As every gardener knows, for new plants to establish on bare soil, it's essential to break up a sealed surface. Animal hooves are adept soil busters. And as more plants establish and grow freely, some dry out before they're eaten. Bunched animals trample the dried material, which then covers the bare soil, making rain more effective by slowing run-off, allowing more water to soak into and remain in the soil by providing cover to reduce evaporation from the soil. More effective rain yields more grass, more litter, and more stored water. It's a natural feedback loop that builds momentum with time, particularly because the increasing effectiveness of the rainfall — not only its volume in absolute numbers but also its effectiveness in penetrating the soil — makes the environment increasingly drought resistant.

Grassland can deteriorate to the point that not a single perennial grass remains. (Perennial grasses provide the greatest stability; grasses and other plants that establish, grow, and die every year have lesser root development and might not grow at all in poor years.) This was the case on the land in the Liebig Advanced Project, where these principles were thoroughly field-tested and the perennial grasses not only returned but also regained their dominance. I saw the same thing happen in the 1960s and '70s, when I was advising owners of ranches in the Karoo area of South Africa; perennial grasses had disappeared, and sheep were subsisting on nothing but desert shrubs. On these ranches, we brought back the perennial grasses without eliminating the shrubs. The ranches now sustain many more sheep and are running cattle as well.

Occasionally, planned grazing begins on land where there is absolutely no plant life at all, as is the case when reclaiming mine dumps. In this scenario, feed (hay) is used to kick-start the biological processes and seedling establishment. Hay is fed on the bare ground and what is not consumed provides the first soil-covering litter. The grazing plan then ensures the first plant seedlings have enough time to grow and survive.

Early Lessons

Perhaps the most powerful lesson to emerge from both Advanced Projects was the recognition that a planning process that accounts for the full complexity of the situation rather than a simple rotation of live-stock is the most critical factor in reviving grassland. After eight years of increasing success, in both good and bad rainfall years, the managers on both sites dropped the grazing planning, and the land suffered. During Zimbabwe's 14-year civil war, which ended in 1979, I was forced into exile because of my political opposition to our racist government. I was unable to return until 1984, by which time both projects had collapsed.

The managers attributed the failure to drought, although there had been no drought. I pointed out to them that almost every plant showed indications of overgrazing, leading to loss of ground cover, and, after all, drought doesn't overgraze plants — only animals do. The managers believed their situation so predictable — relatively flat ground, similar land divisions, one herd of cattle — that they had assumed they no longer needed to do the holistic planning (which involved only about two hours of work every year) and dropped it. Instead, they had reverted to simply moving the herd every day or two to an adjacent paddock, while keeping an eye on grass growth and cattle condition. What they couldn't see was that the animals were returning to graze plants before those plants had recovered from the previous grazing, something the planning charts would have immediately highlighted.

This subtle issue — a reversion to merely rotational grazing — is a common cause of failure on ranches that had started practicing holistic planned grazing. In these scenarios, ranchers who have a preference for working in the field rather than spending even the short time necessary to think and plan on paper (whether they are planning grazing or finances) tend to fly by the seat of their pants, as it were. They drop the planning and the planning chart, believing they can make do by simply bunching the animals and varying the grazing periods based on whether the animals seem hungry or the grass is growing slowly. What they lose sight of is, when you reduce the grazing time in one paddock, you automatically reduce the recovery time in every paddock. As a result, the livestock end up returning to paddocks before the previously grazed plants have fully regrown their roots, and many plants are overgrazed.

The land on Liebig's ranch four years after the grazing planning was dropped.

Scientists and land managers have long tried to create the ideal grazing system in hopes of designing a one-size-fits-all protocol. But the reality is, no prescribed grazing system, including rotational grazing, no matter how flexible or adaptive, can address the complexity that must be addressed when dealing with nature, people and finances. Only a continuous and dynamic planning process can do that. This has led to many years of argument and skepticism of planned grazing from scientists trained to do research that requires replication. The two fields — research using replicable experiments and Holistic Management — are very different. The former is reductionist (a technical term that refers to the isolation of variables for testing), the latter holistic. So where I have claimed that we can, and are, addressing the full complexity (social, environmental, and economic) of the situation on a ranch through holistic planning, researchers have repeatedly dropped the planning process because they cannot replicate it. They have reverted to studying various forms of basic, short-duration, high density, mob or rotational grazing – all of which are possible to replicate because they involve nothing more variable than rotation on a prescribed, regular basis with no accommodation for other vital management considerations. This has led these scientists to conclude that, because their simplified replications failed, my claims for holistic decision making and planning of grazing have been false. Further, they have insisted that any successes are anecdotal because they were not derived from replicated experiment. In short: Because they can't replicate a type of management that by its very nature is unique to every situation managed, researchers trained in the experimental process have doubted my work.

This is a research challenge, one that scientists need to address. But this should not come at the expense of holding up holistic solutions to range management when so many are suffering and dying as a result of global desertification. That management needs to be holistic, embracing all scientific and other sources of knowledge, and can't ever be reductionist — is something I believe no scientist should take issue with.

In any business, management is greatly facilitated when you establish procedures, or management "systems," for handling the routine tasks that generally have predictable outcomes — tracking inventory, managing payables and receivables, maintaining equipment, etc. However, because markets, materials costs, customers, clients, regulations, and taxation are constantly in flux, a business will fail if it sticks to any prescribed system for the overall management of the business, no matter how flexible. Smart business leaders know this and it's why they opt instead for continual planning — and replanning — to manage their companies effectively. Managing any farm or ranch involving livestock grazing is no different: It must account for the complexity inherent in managing nature as well as the social and economic factors that might influence outcomes.

The word "holistic" came into the grazing planning process when ranchers started running into financial and social challenges. Several early planned grazing efforts achieved amazing results on the land, but the ranchers still went broke. For instance, some farmers had accepted low-interest government loans to install irrigation systems and became heavily indebted as the amount of machinery, pumping costs, and so on increased. Having run my own irrigated farm, as well as a ranch, I tried to warn them of the hidden costs of the borrowed money. But my words often fell on deaf ears. After these ranchers went bankrupt, the "newfangled grazing scheme" they had adopted was sometimes held up as the cause. But that was my point all along — addressing livestock handling, even in all its complexity (presence of wildlife, erratic seasons, different soils, drier than normal years), isn't enough. Factors outside this framework — social, economic, environmental, as elaborated above — have to be taken into account. From that point, I decided not to assist any farmer or rancher unless we could deal with the entire business, not just the grazing planning.

In 1966, the United Nations subjected Rhodesia to international economic sanctions to bring down the white government that had declared independence from British rule, precipitating our long civil war. (I later served in Parliament as leader of the opposition to this government.) The sanctions strained many farmers and ranchers, and some ranches began to fail. Every vacant farm or ranch in effect became a potential guerrilla base. To avoid this, the government helped prop up failing farmers. The Rhodesian government formed a financing entity, known as the AFC (Agricultural Finance Corporation) that gave farmers money for essentials, such as school fees and farming supplies, so they could remain on the land and continue farming. The AFC engaged me to work with these bankrupt farmers to assist them in getting back on their feet. I soon found that as long as a farmer was not mentally defeated, we could turn things around. It taught me that you could never manage land without accounting for the attitudes of the managers, their families, and their finances. This is why holistic management today treats the land, finances, and people involved as inseparable. Each area feeds into the others, which makes the overall result ecologically, economically, and financially successful simultaneously.

Impressive Results

Holistic planned grazing is now practiced on more than 14 million hectares across six continents with consistently encouraging results. The following images illustrate the results in different grassland environments.

U.S.A.

Wyoming: Photos were taken from the same bridge moments apart. Upstream land (left) was under holistic planned grazing and a 250 percent increase in livestock numbers for about 20 years. Downstream land (right) was grazed conventionally.
Image: Andrea Malmberg

MEXICO

Guillermo Osuna, owner of Las Pilas Ranch, describes the photos:

The pictures were taken in 1953 and 2007. The

drought of the 1950s was certainly a factor, but the dramatic change we can see in the pictures was due to improving the ground cover:

The changes began when I first realized that, as cattlemen, our first objective should be to grow grass, which we could then harvest with livestock. Real progress in this direction really began in 1980, though, when we first started practicing what we now know as Holistic Planned Grazing. It is interesting to note that the average rainfall for the last six decades here at Las Pilas has really not changed very much, although the year-to-year fluctuations can be very big. In 2010 we had the biggest rainfall of record, 64 inches, and in 2011 we got less than 4 inches.

Coahuila, Mexico: Las Pilas Ranch, with a man-made pond dug to provide water for scattered livestock (top) and after 25 years of Holistic Planned Grazing (bottom). The arrow indicates the same hill in both photos.
Image: Courtesy of Guillermo Osuna

ZIMBABWE

HERDING VERSUS FENCING

Instead of using fences to control animals, humans can herd them. We've been doing this in Zimbabwe since 2002 on a ranch owned by the Africa Centre for Holistic Management (ACHM), an organization I co-founded in 1992. On a large scale, we've found that herding beats fencing, in terms of speed of land recovery. With fencing, there are costs to erect and maintain it, and labor costs thereafter are low.

A river gone dry due to desertification has led a formerly self-reliant people to become dependent on food aid.

A river on nearby land with the same soils, vegetation, and rainfall and seen on the same day, but managed under holistic planned grazing, looks quite different.

However, the animals are scattered within the fenced area while grazing and trampling — leading to a high level of partial rest for grasses and plants, a central cause of desertification and the most difficult to deal with on a vast scale over large areas of land. Herding also beats fencing when you are dealing with large wildlife populations, as fencing is so destructive to them and so expensive to maintain when you have elephants, buffalo and other large herbivores continually breaking through it. (These problems vanish when you create virtual, unfenced, paddocks and herd within them.)

With herding, although people or dogs are needed to supervise, the animals are always more concentrated, so the land is consistently subjected to more animal impact, which breaks sealed soil surfaces and lays down more ground-covering dead plant litter. Because grass regeneration benefits from animal bunching of this kind, more animals can be run, and thus costs are spread over more animals, leading not only to greater productivity per acre but potentially greater prosperity, which, in turn, means increased profit, more employment, more money recycled into the community — the model of why such decisions have to be made holistically. Where predation is not an issue, animals can simply be positioned and left overnight to resume herding the next day. But where predation is an issue, and because predators benefit the wildlife (and ultimately the land) by keeping animals moving, we confine the animals overnight in movable kraals (corrals) — curtains of durable, sun-resistant plastic that are easily moved every week or so. These overnight enclosures can be placed on the worst land to speed its regeneration.

To appreciate the difference between herding and fencing, consider what I witnessed on two ranches where I was involved in coaching the management, one in West Texas and the other the ACHM ranch itself in Zimbabwe. Both ranches were similar in size, in distribution of humidity, and in livestock numbers and composition. The Texas ranch controlled livestock with fencing. In Zimbabwe, we delineated grazing areas by land features, such as roads, streams, hills, etc., and then herded the livestock within these "imaginary" paddocks.

On the Texas ranch, the animals were mostly

scattered, even though they were in fenced paddocks. In Zimbabwe, the same number of "paddocks" existed for planning, but herders kept the animals bunched during the day, and then at night contained them in movable kraals to protect them from lions.

On the Texas ranch, plants were grazed (not over-grazed) and the "animal impact" (trampling, dunging, urinating) increased, though overall the land was still partially rested. In Zimbabwe, plants were also grazed (not overgrazed), but the animal impact was many times greater, overresting greatly minimized, and the land improved much faster. So much so that in 2013, following the lowest rainfall in many years (about half the average rainfall), we grew more grass than we ever did in the best years in the past. And we are doubling the animal numbers simply to keep pace with the production of the land, now that the available rainfall is so much more effective. The flexibility that herding provides to the manager far surpasses that of fencing over large land areas. However, herding or fencing, as any other decision, will have to be checked by the decision makers to make sure it is aligned with their holistic context.

This is good news for pastoralists who herd live-stock in some of the most problematic regions of the world. Many pastoralists are being forcibly moved off their traditional lands and resettled in villages where livestock are kept in pens and fed farmed forage. Herding under Holistic Planned Grazing means these people can begin restoring their land and reverse the cultural genocide. Pastoralists in the violent Horn of Africa have said this is the only hope for saving their families and culture.

SOUTH AFRICA

The Kroon family ranch is in the Karoo area of South Africa. The Kroons were one of my earliest clients. The left side of this 2012 photo demonstrates what livestock managed under holistic planned grazing have done to reclaim land turning to desert.
Image: Norman Kroon

CHILE

A common question I'm often asked: How long does it take to get results after changing to Holistic Planned Grazing? The answer: within the first year. In Patagonia (Chile) flocks as large as 25,000 sheep under planned grazing produced a measured 50 percent improvement in the land in the first year. In general, the larger the herd size, the greater the bunching and impact achieved, and the faster the recovery — although good rains always help!

Patagonia: Herders move a flock of 25,000 sheep from one grazing area to another.
Image: Courtesy of Jose Manuel Gortazar

Remember: Overresting land is the leading cause of desertification in seasonally humid grasslands, and only animal impact provided by bunched animals can address it.

COMMUNAL LANDS

Holistic Planned Grazing is being adopted in communal lands, where communities are combining individually owned animals into "land management herds." Literacy rates in these communities are low. But because their powers of observation and memory skills are often very high, we've been able to simplify the planning process and planning chart to meet their needs. We're now training villagers and pastoralists in southern Africa to gather livestock into communal herds and plan their grazing holistically.

THE CASE FOR KRAALS

With funding from the U.S. Agency for International Development (USAID) Office of Foreign Disaster Assistance and other organizations, many communities in Southern Africa are implementing Holistic Planned Grazing. Livestock in these areas move in

conjunction with many wild herbivores, and predators remain important for keeping wild herds bunched
and moving. Hence, it's important to run livestock in
a predator-friendly manner. To do so, herders move
livestock according to plan and then remain present
at all times to ward off predators — lions, wild dogs,
cheetahs, leopards, and hyenas. At night, livestock
are retained in kraals (corrals). Communities that for
centuries had permanent overnight kraals are adopt-
ing movable kraals designed by the Africa Centre for
Holistic Management (ACHM).

Portable kraals vary in area depending on the herd size. On ACHM's ranch there might be 500 to 600 cattle in a one-acre kraal. At night, the kraal site concentrates the trampling, dunging, and urinating. Dung and urine are assets when scat-tered on land but become

This movable kraal on the Africa Centre for Holistic Management ranch in Zimbabwe is a promising tool for Holistic Planned Grazing.

pollutants when concentrated too heavily in any one
place. For this reason, kraals are moved about once a
week. We also use
kraals to more rapidly heal bare and compacted areas
or eroding gullies by placing them over these areas to
give them a high dose of trampling, dung, and urine.

Moreover, the movable kraals are leading to great
increases in maize yields — two times higher, and in

some cases as much as five — when they're placed on community crop fields prior to planting.

The maize on the right shows the difference in yield where, six months earlier, a kraal had been placed for a week and the livestock trampled the soil.

And there is no need to plow the fields, nor cart manure, nor purchase fertilizers. This is not a new idea — it has been used for centuries in some parts of India, where crop farmers make agreements with pastoralists to overnight their animals in small movable pens on harvested fields.

Chapter 7

Scaling Up

In May 2013, the concentration of carbon dioxide in the atmosphere surpassed 400 parts per million (ppm) for the first time in at least 3 million years. If the upward trend continues, the Intergovernmental Panel on Climate Change predicts carbon dioxide concentrations could rise above 800 ppm by the end of this century — with devastating consequences. Even the best-case scenario, which assumes "rapid changes in economic structures" that favor reducing global emissions, would only bring that number to roughly 500 ppm. And that's why efforts to reduce emissions aren't enough. We must decrease the excess carbon already in the atmosphere — referred to as the "legacy load" — to a concentration of no more than 350 ppm, "if humanity wishes to preserve a planet similar to that on which civilization developed and to which life on Earth is adapted," warns Jim Hansen, the NASA scientist who first sounded the alarm on climate change nearly three decades ago.[9]

To decrease the legacy load in the atmosphere, soil scientists have investigated the potential for storing carbon in soils suitable for crop growing, which constitute roughly 30 percent of the world's land area. But even the best conservation agriculture practices, such as no-till and cover crops, won't reduce the legacy load enough.[10] Scientists disagree on just how much carbon grasslands can sequester, but if, indeed, grasslands can sequester carbon on the scale I believe them capable of and what more recently has been reported in peer-reviewed papers (in the range of 3 to 7 tons of carbon per hectare per year)[11] reversing desertification may be essential to returning the planet to 350 ppm.

Putting climate change aside for a moment, consider that desertification contributes significantly to poverty and violence in the world's most distressed regions. Here, millions of people are already running out of water, forage for their animals, and food for themselves due to increasingly frequent man-made droughts, which in turn leads to violence and even war, and to the suffering and deaths of millions. An average of only 5 percent of the land area in these regions is capable of growing crops to feed people. The bulk of the rest is suitable only for livestock production, because they are grasslands, which require less rainfall and less reliable rainfall. And it is only livestock, using the holistic grazing planning process, or better when developed, that can begin to reverse the desertification of these grasslands and the problems arising from it.

As the pace of desertification slows and begins to reverse, soils begin to store more water. Soil is, as mentioned earlier, the largest reservoir of non-frozen fresh water available. If, for instance, we were to make just 1 inch (25 mm) of the rain that falls in severely desertifying New Mexico each year soak into the soil rather than evaporate or run off, more water would be stored in the soil every year than in three reservoirs the size of New Mexico's largest reservoir.[12]

Bear in mind that this extra water remaining in the soil is added to with each following year in a cumulative manner. And bear in mind that the fate of both carbon and water in the soil is tied to increasing organic matter and diversity of life both in and above the soil surface.

In the United States, millions of cattle are fed grain in a fossil-fuel-based factory production system, while so much of the land in the western half of the country is desertifying *due to too few livestock*. Should this continue, it won't be long before American ranching culture will be remembered only at cowboy poetry festivals, rodeos, or in Hollywood films.

We must change our policies and practices now because we know that holistic planned grazing can:

- Help heal human-made deserts, which will once again assume their role as productive grasslands;

- Restore desertified grasslands to mitigate floods, droughts, stabilize river flow, and greatly increase water storage;
- Let ranchers increase livestock (and other herbivore) numbers to hasten land recovery, no matter how bad its condition;
- Dramatically increase wildlife, soil, and plant diversity;
- Reduce carbon that is currently being added to the atmosphere by dying eroding soils, while sequestering excess atmospheric carbon through regenerating grassland soils that cover enormous areas of the earth's land surface;
- Address a major underlying cause of human-made floods, droughts, poverty, social breakdown, emigration, cultural genocide, violence, and war; and,
- Improve community food supplies and financial security.

The possibilities are indeed endless once we begin to reverse desertification in the grasslands — and former grasslands — of the world.

My early blunder as a young scientist led to the futile deaths of more than 40,000 elephants. It was an epic mistake that truly changed me. Had I known then what I know now, I might have been able to save the elephants and establish long-term policies that better served the people and wildlife of Zimbabwe. Today, I live half the year near Victoria Falls, in Zimbabwe, where many elephants reside in and around the national parks. And yet, these grand giants remain scapegoats for much of the parkland desertification evident in southern Africa. I have worked with, or visited, wonderful, proud pastoral people in Africa, Pakistan, and Israel. But their governments and assisting international organizations embrace outdated thinking, blaming livestock for their land's afflictions and calling for their removal.

Nonetheless, managing holistically – involving, as it does, science and other sources of knowledge – is backed by solid data. Scientists at Ohio State University conducted a survey of 25 holistically managed properties across the U.S. They examined the land — its health and productivity — and on all but one ranch found improved biodiversity, while profitability (which had increased in four-fifths of cases) had on average, in those cases, nearly quadrupled (the median profit increase was 238 percent).[13] Even so, I know of no scientific criticism of our holistic management framework that addresses its full social, cultural, economic, and environmental complexity.

Detractors typically focus on components of the process but never consider the whole package. Meanwhile, as of 2022 we have trained more than 15,000 people — not only ranchers and farmers, but also scientists from the USDA, World Bank, USAID, U.S. Fish and Wildlife Service, and faculty members from U.S. land grant colleges — and I've appealed to them all to help find the flaws in the principles and practices involved in managing holistically. Of course, it would be abnormal if nobody questioned my methods. Such is the fate of any new, counter-intuitive, paradigm-shifting advance.

After many years of struggle against institutional resistance that even people leading those institutions could not change, it is my opinion that institutions, even if they wish to, cannot change until there is a significant sway of public opinion. I base this opinion on my own experience in trying to lead genuine change as president of a political party, the centuries it took the Catholic Church to accept that the earth was not the center of the universe, the 200 years it took the Royal Navy to accept that lime juice could prevent scurvy, and the research of Eric Ashby about the need for a certain level of shift in society's view before people in elected positions in democratic institutions, or the institutions themselves, could change.

The global problem of desertification with all its many adverse effects cannot be realistically addressed on the scale needed until there is institutional change. It's my hope that I can begin that shift here. Thanks to TED, my ideas have reached millions of people for the first time. Even prior to TED, local branches of organizations such as The Nature Conservancy were showing new openness to holistic management and have been collaborating on the development and implementation of grazing plans with the Savory Institute, which I co-founded with five others to scale holistic planned grazing globally. Our strategy is two-pronged:

1. Increase adoption of Holistic Management with the creation of locally owned and operated Savory Hubs that, through Accredited Professionals, provide Holistic Management training, resources, and implementation support to local ranchers, farmers, and pastoralists.
2. Remove barriers to large-scale change through:
    - Monitoring, curating, and conducting research with a variety of collaborators to document the environmental, economic, and social changes that occur as a result of managing livestock holistically;
    - Informing policies and programs related to grasslands and livestock management of reasons why desertification is spreading rapidly and how it can be reversed with Holistic Planned Grazing;
    - Identifying and establishing market-based incentives that encourage people to manage holistically;
    - Increasing public awareness that livestock are not the problem. It is how they are managed and interact with grasslands that is the problem. And this is an issue of human management and decision making.

In the decade following my TED Talk, we have made tremendous headway:

- Established 54 regional learning Hubs and accredited over 200 professional educators around the globe;
- Trained over 15,000 farmers, ranchers, and pastoralists and influenced management of nearly 22 million hectares of grasslands through the adoption of Holistic Planned Grazing;
- Collaborated with a team of scientists on the creation of a protocol for verifying ecological outcomes, and trained and accredited a rapidly growing number of monitors, plus hub verifiers and master verifiers who ensure the quality of the data monitors gather (over a million hectares are enrolled in the Ecological Outcomes Verification, or EOV, program and that figure is growing rapidly);
- Launched Land to Market, a program that was spun off in 2022 as a for-profit company that brings verified regenerative (through EOV) producers together with brands seeking to source their meat, wool, dairy, and leather supplies from livestock raised on land that is verified to be regenerating;
- Some 24 peer-reviewed papers have been published on the positive social, environmental, and economic outcomes where people are planning their grazing holistically; and,

· SI now serves as a member or adviser in several global roundtables and dialogues formed to influence agricultural and environmental policies and programs.

And, of course, we hope to further engage the millions of people who have watched my TED Talk or read this book so the change in public opinion advances ever more rapidly.

Thanks

While I have been the person driving the development of a solution to the riddle of desertification, I have not worked alone. I am indebted to the many hundreds of ranchers, farmers, pastoralists, scientists, and others who tested the ideas and provided such valuable criticism and thoughtful feedback.

I owe a great deal to the handful of individuals who co-founded the Savory Institute in 2009 when the gains we had made so painfully over so many years were at risk of being lost. In particular I thank my wife of over four decades, Jody Butterfield, who is also my editor and collaborator and Daniela Ibarra-Howell, one of my early students and now Savory Institute's CEO, whose determination and never-ending support sustained me in some of the most challenging times. For this brief book based on my 2013 TED Talk, which has now been viewed by some 8 million people, and still by an average of 2,000 viewers per day, I have to thank TED, who first proposed it.

References

1. J. Carey, "Global Warming: Faster than Expected?," *Scientific American* 30, no. 5 (November 2012): 50–55.
2. N. Myers, *Gaia: An Atlas of Planet Management* (Garden City, NY: Anchor/Doubleday, 1993).
3. W.C. Lowdermilk, "Conquest of the Land through 7,000 Years," *USDA Bulletin* 99 (Washington, DC: U.S. Department of Agriculture, Natural Resources Conservation Service, 1939).
4. M. Seely et al., *Integrated Land and Water Management in Africa: Napcod — Working Towards Communities Taking the Lead: A Case Study from Namibia* (Location: Global Environmental Facility STAP Expert Group, 2002).
5. *Desertification of the United States* (Washington, DC: Council on Environmental Quality, 1981), 334–983.
6. A. Voisin, *Grass Productivity*, Second Edition (Washington, D.C.: Island Press, 1988).
7. J.E. Russell, *Soil Conditions and Plant Growth*, Ninth Edition (London: Longmans, 1961).
8. G. Mossop, *Running the Gauntlet: Some Recollections of Adventure*, Second Edition (Pietermaritzburg, South Africa: G.C. Burton, 1990): 7.
9. J. Hansen et al., "Target Atmospheric CO2: Where Should Humanity Aim?, *Open Atmos. Sci. J.* 2 (2008): 217–231.
10. R. Lal, "Soil Management and Restoration for C Seques-

tration to Mitigate the Accelerated Greenhouse Effect," *Progress in Environment Science* 1, no. 4 (1999): 307– 326.

11. W R Teague, "Forages and Pastures Symposium: Cover Crops in Livestock Production: Whole-System Approach: Managing grazing to restore soil health and farm livelihoods," *Journal of Animal Science* 96, no.4 (April, 2018): 1519–1530.

12. The calculation for this example was derived by the following method (with thanks to Keith Weber of Idaho State University):

    1 mm rainfall on 1 km2 = 1,000 m3

    0.001 meter x 1,000,000 m2 = 10,000 m3

    (A sq. km is 1,000 m x 1,000 m or 1 million sq. meters in area.) New Mexico is 315,000 km in area.

    The volume (in cubic meters [m3]) of water striking New Mexico during a 1 mm rainfall event is then: 315,000 x 10,000 = 3,150,000,000 m3 (A)

    Since 1 inch = 25 mm, we now need to multiply A by 25, which gives us 78,750,000,000 m3 of water (B)

    B converted to km3 = 7.8 km3

    Since Elephant Butte reservoir contains 2.7 km3 of water, the amount of water in the example is equal to 7.8/2.7 = 2.9, or about three reservoirs the size of Elephant Butte reservoir.

13. D.H. Stinner et al., "Biodiversity as an Organizing Principle in Agroecosystem Management: Case Studies of Holistic Resource Management Practitioners in the USA," *Agriculture, Ecosystems and Environment* 62 (1997): 199–213.

About the Author

Allan Savory was born in Zimbabwe and educated in South Africa (University of Natal – BSc Biology and Botany). Since the 1950s he has been devoted to understanding the causes behind the degradation and desertification of the world's grassland environments.

After emigrating to the United States, he co-founded the Savory Institute to promote large-scale restoration of the world's grasslands through Holistic Management. In 1992 he co-founded the Africa Centre for Holistic Management near Victoria Falls, Zimbabwe to provide training to organizations working with pastoral communities to restore degraded lands and water sources to health.

Savory's book, now in its third edition, *Holistic Management: A Commonsense Revolution to Restore Our Environment* (Island Press, 2016), describes his effort to find workable solutions ordinary people could implement to overcome many of the problems besetting communities, businesses and governments today.

About the Savory Institute

Since 2009, the Savory Institute has been leading the regenerative agriculture movement as a nonprofit dedicated to helping farmers, ranchers, and pastoralist communities regenerate their grasslands. Using the Holistic Management framework for managing complexity and the Ecological Outcome Verification (EOV) protocol for assessing soil health, biodiversity, and ecosystem function, Savory Institute's global network of more than 50 regional Hubs creates localized impact solutions that build verifiable regenerative results. As of 2022, Savory Institute has influenced more than 54 million acres (21 million hectares) around the world and equipped 15,000+ farmers and ranchers with education, training, and implementation support within their own holistic contexts ensuring they manage their lives, businesses, economy and life-supporting environment inseparably.

Contact us at:

885 Arapahoe Ave
Boulder CO 80302
www.savory.global, info@savory.global

See our YouTube channel for short videos illustrating many of the points covered in this book.